WAW

Choosing a Career in the Military

Military life can be demanding but also fulfilling.

Choosing a Career in the Military

Greg Roza

The Rosen Publishing Group, Inc.
New York

In memory of William McConnell

Published in 2001 by The Rosen Publishing Group, Inc.
29 East 21st Street, New York, NY 10010

First Edition

Library of Congress Cataloging-in-Publication Data

Roza, Greg.
  Careers in the military / Greg Roza—1st ed.
      p.   cm. — (World of work)
  Includes bibliographical references and index.
  ISBN: 978-1-4358-8690-2
  1. United States Armed Forces—Vocational guidance—Juvenile
literature. [1. United States Armed Forces—Vocational guidance. 2.
Vocational Guidance.] I. Title. II. World of work (New York, N.Y.)
  UB147.R69   2000
  355'. 0023'73—dc21   00-010217

Manufactured in the United States of America

Contents

Introduction

*C*harles was seventeen and a senior in high school. Graduation was quickly approaching, and it excited him to think that soon he would be entering "the real world." Charles was afraid he would end up working at a fast-food restaurant since he didn't have a job lined up. He was a good mechanic, but so what? He had thought about going to college, but he decided that he wanted to get away from school for a while and see more of the world.

Charles's cousin, Sue, showed up unexpectedly at his graduation. She had been away for a year, training to become an electrician with the United States Army. She told him that she was on a two-week leave, and then she would be traveling to an American base in Germany where she would get the chance to practice her skills and see some of the world. "Wow," Charles

*thought. "Sue sure is lucky. I wonder if
I should join the army. I wonder if the
army could use a good mechanic."*

Young people like Charles face situations like
this all the time. While college might be the answer
for many high school graduates, it isn't the only
option. Choosing a career in the military might be
the perfect choice for you.

Military life can be tough at times. It is a
demanding environment requiring determination,
a strong desire to succeed, and a passionate sense
of pride in your nation. A military career can also
be a highly fulfilling lifestyle. Enlisting in the
military is a good way for you to discover what you
are good at. Many people make a lifelong career in
the military, but that is not necessarily for everyone.
You can sign up for two, three, four, or more years,
and then take the knowledge you have learned with
you into the civilian world.

Whatever your talents or interests might be, a
career in the military can prepare you for what
you want to do in life. It can also give you the
opportunity to travel the world, engage in exciting
activities, and meet interesting people. Let's read
on and explore the branches of the military to find
out if a career in the military might be best for you.

The military, the largest employer in the nation, offers enlistees the potential for a long and rewarding career serving their country.

1

Preparing for the Military

*J*oe was a junior in high school. His father had enjoyed a long, successful career in the United States Navy and had recently retired to run a family business. Joe's grades were average, but with a little hard work he might be guaranteed placement at a state university. However, Joe had always thought about following in his father's footsteps.

One day Joe said, "Dad, I was thinking about joining the navy."

Joe's father's eyes lit up. "That is a great idea, Joe. Have you thought about taking the ASVAB this summer?"

"ASVAB? What's that?" Joe asked.

"The Armed Services Vocational Aptitude Battery," his father responded. "All enlistees have to take this test before entering the service, and there's no reason you couldn't get a jump start on this if you're really serious."

Joe's father was right. For those of you who have a year or two to go until you graduate high school, it isn't too early to think about starting the enlistment process. Of course, enlisting in the military is not a decision to be taken lightly. There are many factors to consider when opting for a military career, and there are several outlets for you to explore to help you make up your mind.

General Facts About the Military

Before addressing the steps you need to take when joining a branch of the military, it might help to know more about military careers in general. When joining the military, enlistees are signing a contract. The military agrees to provide you with a steady job, good wages, benefits, and training in any of a large number of occupations. In return, the enlistee agrees to serve for a specified period of time, which is called a service obligation. The standard service obligation—which may differ from one branch of the military to the other—is approximately eight years. Enlistees must serve at least two years on active duty, or full-time military duty. The rest of the contract may be fulfilled in the reserves (further discussed in chapter 3).

The pay scale in the military, set by Congress, is the same in all branches. Currently, all enlistees earn $959.40 a month upon entering the service and earn cost-of-living increases every year that they serve. In addition, food, clothing, and shelter are provided. Incentive pay is given to enlistees engaged in special or hazardous work, such as parachute jumping and explosives demolition.

The military respects and cares for its workers and offers a generous package of benefits and bonuses, including medical and dental care, vacation time (one month for every year you serve), free legal assistance, recreational programs, and retirement plans. Particularly of interest for young enlistees might be the unbeatable educational assistance provided by the military. Upon signing the Montgomery GI Bill (MGIB), you agree to contribute $100 a month from your salary for twelve months. In return, you earn up to $19,000 toward a college education after you complete active duty.

The military is the largest employer in the nation. Nearly 1.5 million active duty individuals are employed by the five branches of the military at any given time. The military offers over 2,000 enlisted job specialties, which are conveniently divided into twelve broad categories:

✔ Human Services
✔ Media and Public Affairs
✔ Health Care
✔ Engineering, Science, and Technology
✔ Administration
✔ Service
✔ Vehicle and Machinery Mechanic
✔ Electronics and Electrical Equipment Repair
✔ Construction
✔ Machine Operator and Precision Work
✔ Transportation and Material Handling
✔ Combat Specialty

The first step toward enlisting in the military is to talk to a recruiting officer.

Training in any of these fields can lead to steady careers in the civilian world as well. We will take a closer look at these categories in chapter 4.

How Early Can I Enlist?

You can enlist in the armed services as early as age seventeen. Although you do not necessarily have to have a high school diploma, military representatives will tell you that it is preferred. While you are finishing your senior year of high school, and after you have thoroughly researched the branch of the military you wish to enter, you can enlist through the Delayed Entry Program (DEP). The DEP allows enlistees to delay their entry into the service up to one year, allowing them to finish high school. (The DEP can also give enlistees time if the specific type of training they are seeking is currently unavailable.) Enlistees under the DEP make plans with their recruiting officers, and upon graduation from high school, enter active duty in the branch of their choice.

The Enlistment Process

When deciding to enlist, you should consult people close to you, as well as someone who has spent time in the military. In addition, it would be wise to research the military and your enlistment options at your local library and on the Internet. It cannot be stressed enough: You should give enlistment into the military careful consideration, since it is a decision that will shape your life.

When you have made the decision to enlist, however, there are several steps you will need to

13

Enlistees undergo rigorous exercise during basic training.

follow to finalize your entrance into the service. The first step is to talk to a recruiting officer. Recruiting officers are available to the public to give information and advice about enlisting. They will inform you about training opportunities and qualification requirements. They will also schedule enlistment processing for you if they see no potential difficulties that may stand in your way (health problems, for example). You should trust your local recruiter to help you decide what is best for your future in the armed services; that is what they are there for. More information on recruitment is in the For More Information section at the back of this book.

Next, you must qualify for enlistment. At this time, applicants must go to a military entrance processing station (MEPS) to take the test Joe's father mentioned, the ASVAB, which helps determine if applicants qualify for military service. ASVAB scores are good for two years, and the test can be taken in high school in preparation for enlistment. Applicants will also receive a medical exam at this time. Then, applicants meet with a service classifier. A service classifier is a military career information specialist. By entering your aptitude scores into a computer that matches them with your interests, the service classifier will figure out which positions are best for you. Then you select an occupation and schedule an enlistment date. Before finalizing this step, you must sign a contract with the branch of your choice, and take the Military Oath of Enlistment, which is the same for all the branches.

The last step in the enlistment process is promptly going to your assigned base for basic training. In the case of the DEP, you are expected to report to the base you have been scheduled to attend on the date decided by your service classifier.

Basic Training

The purpose of basic training (lasting six to eleven weeks) is to prepare new enlistees for service in the armed forces, both mentally and physically. Not only will you be required to undergo rigorous physical training, you will also be required to attend and pass classes with subjects relating to your experience in the military. You will be trained in such areas as weapons use, first aid, and map reading, to name a few. Basic training for the five branches differs. The base at which an enlistee trains depends on the job training he or she is to receive. Recruits are divided into groups of forty to eighty people. Soon after these groups are formed, they meet with their drill instructors, receive uniforms and supplies, and move into their assigned quarters.

Daily exercises are designed to improve conditioning, stamina, and overall fitness. Although physical conditioning is a major focus during basic training, equally important is building a sense of pride and discipline. When basic training is completed, enlistees move on to job training, which often takes place in another location.

Reserve Officers Training Corps (ROTC)

If you are torn between attending college and enlisting in a branch of the military, the Reserve Officers Training Corps, or ROTC, might be the answer for you. This program is an elective course offered at many colleges and universities around the country, and is staffed by active-duty personnel. ROTC is the single greatest supplier of officers to the armed forces today, in both active and reserve duty. The army, navy, and air force each run separate ROTC programs. Regular and reserve officers for the Marine Corps come from the navy ROTC program.

ROTC enrollees learn leadership and management skills while actively participating in training schedules similar to those in the military. Sometimes this means spending your Saturday mornings running or doing calisthenics. Sometimes it means white-water rafting while your college classmates are cramming for their next exam. Of course, ROTC students are also enrolled full-time in college. Being involved in an ROTC program means being strongly involved in hands-on training for an officer position in the armed forces.

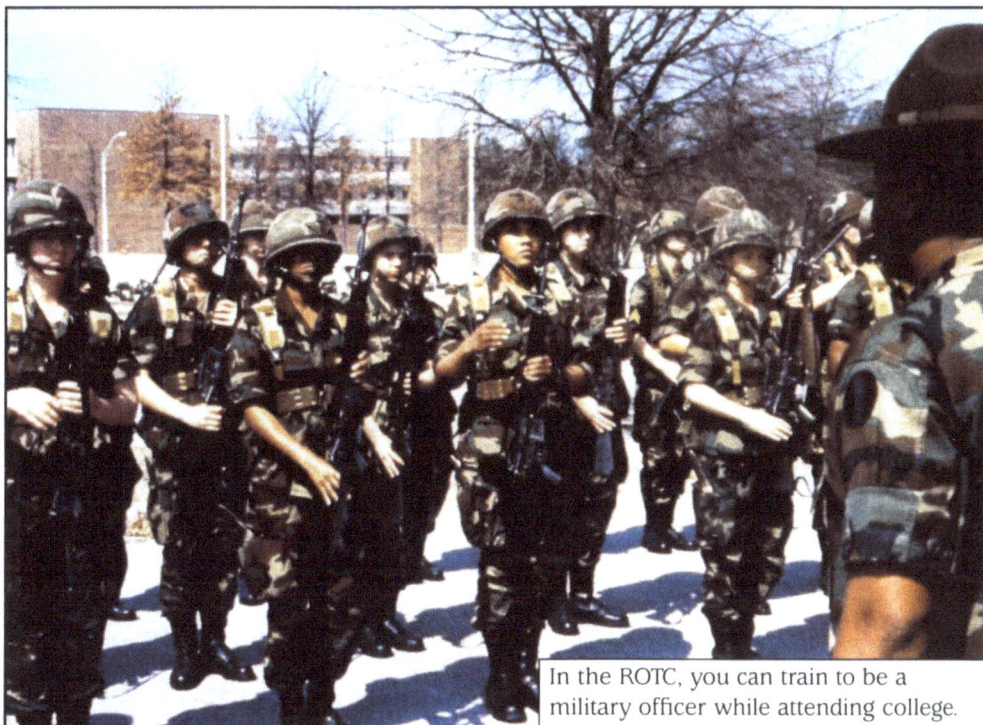
In the ROTC, you can train to be a military officer while attending college.

A typical ROTC program takes four years to complete, fitting in very well with a standard college class load. You can enroll in two years of ROTC training without being obligated to join the armed services upon graduation. Students interested in an ROTC program will encounter a mixture of classroom time and hands-on experience.

A Large Responsibility

In the military you will be expected to follow rules, listen to orders, and work with your peers. You will gain valuable career training, travel the world, and you'll make new and enduring friendships. It should be stated that a military enlistee should have a great deal of pride in being an American citizen—enough pride to help his or her country any way he or she can in times of war, as well as in times of peace.

2

Branches of
the Military

B *oris had taken his ASVAB test in his*
final months of high school. His
scores were above average, but he wasn't
sure which branch of the military he
wanted to join. He kind of wanted to be
a marine, but he had always wanted to
fly a fighter jet. Maybe the air force was
the right choice. Then again, he had
always enjoyed sailing with his grand-
father; maybe he should join the navy.

With so many great options, how
could he choose one? He decided to visit
a few recruiters to get some help. He
needed more information about the five
branches of the military before he could
make up his mind.

The five branches of the armed services—the
army, the navy, the air force, the marines, and the
Coast Guard—each offer many career opportunities
for young enlistees, not to mention competitive wages
and promotion prospects. Deciding which branch is

best for you might be a difficult decision. However, once you have properly researched your options, you should be able to make a well-informed decision. The following section will present each branch in a concise, yet informative way.

Which Branch Is Right for You?

Army

The primary purpose of the United States Army is to fight and win our wars on land. This is accomplished with the aid of all the branches of the military. On average, the army employs 69,000 officers, 11,500 warrant officers, and 450,000 enlisted soldiers. Men and women in the army have many kinds of jobs, ranging from general administration to the operation and maintenance of thousands of weapons, vehicles, aircraft, and state-of-the-art electronic systems.

The army seeks approximately 80,000 to 90,000 new enlistees each year. New enlistees will find hundreds of challenging career opportunities that offer a lifetime of security and excitement. Applicants must be seventeen to thirty-five years old, American citizens or registered aliens, and in good health and physical condition. Army enlistment may be for two, three, four, five, or six years. All applicants must take the Armed Services Vocational Aptitude Battery to determine which careers they are best suited for.

In most cases, applicants are usually assured their choice of career training. For applicants who wish to be guaranteed a specific school, a particular area of assignment, or both, the army offers the

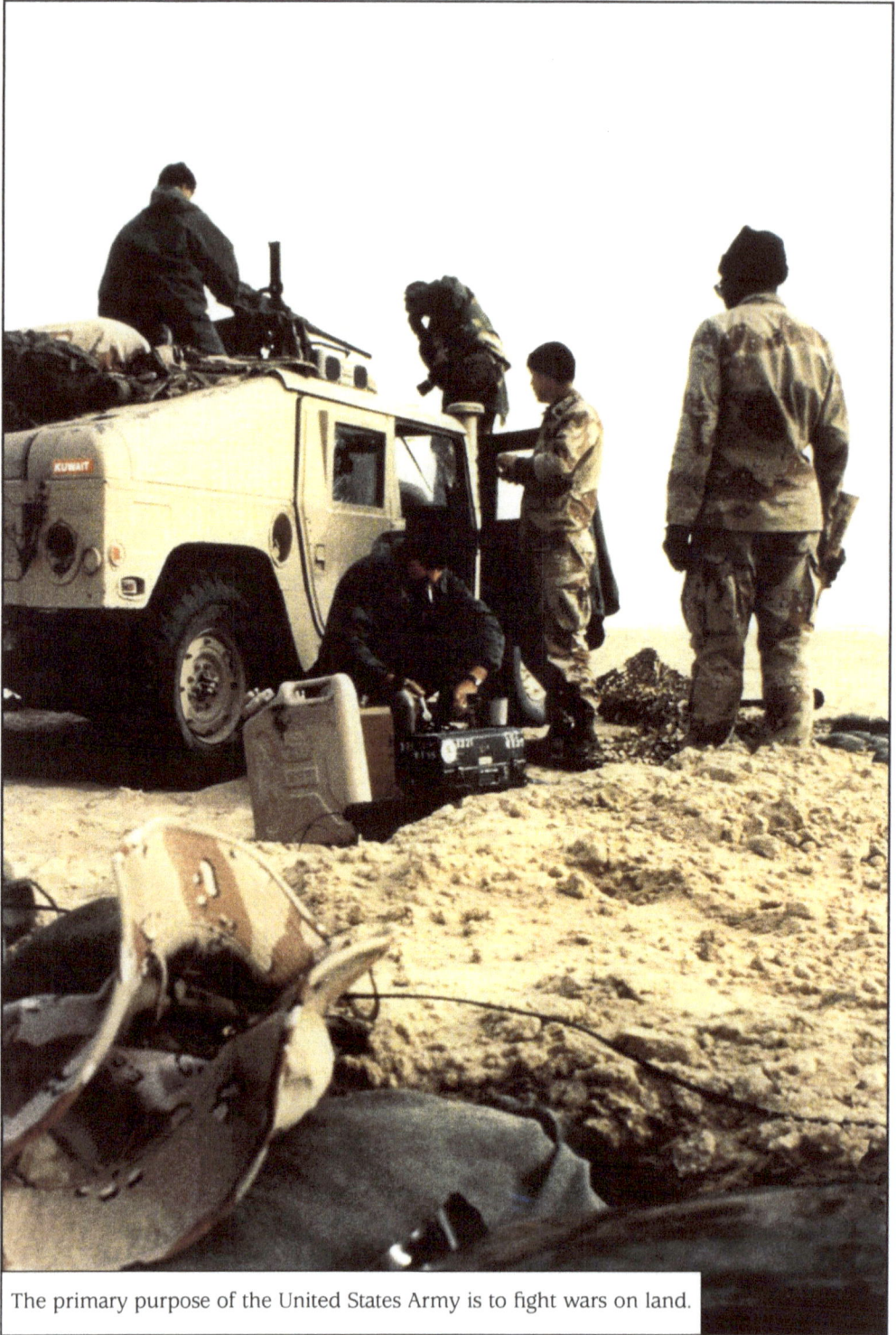

The primary purpose of the United States Army is to fight wars on land.

Delayed Entry Program. DEP applicants can reserve a school or an assignment choice up to one year in advance of entry into the army. In some cases, the army also offers enlistment bonuses.

Navy

The United States Navy plays an important role in maintaining the freedom of the seas. It defends the right of our country and our allies to travel and trade freely on the world's oceans. It also protects our country during times of war. Navy presence guarantees the United States its right to use the oceans whenever our national well-being requires it.

Today's navy is made up of nearly 400,000 active-duty men and women. Approximately 336,000 of these soldiers are enlisted sailors and midshipmen. The navy recruits about 60,000 officers and enlisted people each year to fill openings in navy career fields. With the exception of the Navy SEALs and submarine assignments, all areas are open to women.

Navy personnel operate and restore more than 340 ships and over 4,000 aircraft. Enlisted personnel serve in such exciting fields as communications, computer programming, and electrical engineering, to name a few. Navy people serve on ships, on submarines, on aircraft, and at bases around the world.

Applicants who wish to enlist in navy programs must be between the ages of seventeen and thirty-four; parental consent is required for all seventeen-year-olds. Because of extensive training requirements, the maximum enlistment age is twenty-five in the nuclear field. Enlistees must be

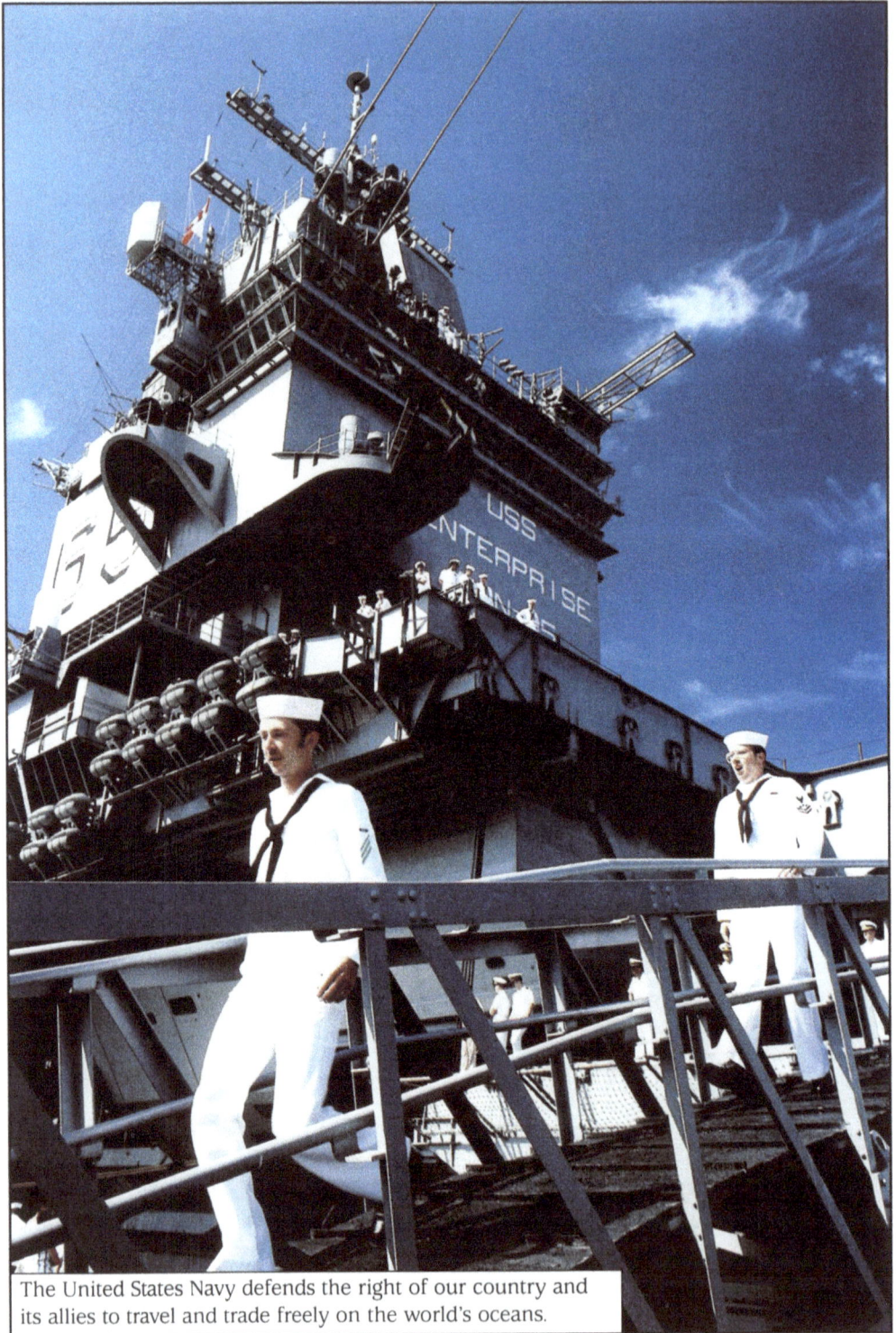

The United States Navy defends the right of our country and
its allies to travel and trade freely on the world's oceans.

citizens of the United States or registered aliens. They must also complete a physical examination and the Armed Services Vocational Aptitude Battery.

The navy provides additional pay for sea duty, submarine duty, demolition duty, diving duty, aviation duty, or jobs that require special training. Many fields, such as the nuclear field, require individuals with unique skills and offer quicker promotions and bonuses when training is completed.

Air Force

The United States Air Force is the aerospace branch of our nation's armed forces. The men and women of the air force maintain and pilot the world's most technically superior aerospace vehicles. These forces are used whenever necessary to protect the interests of the United States and our allies.

The air force is made up of nearly 385,000 men and women. About 76,000 officers man aircraft, launch satellites, gather intelligence data, manage maintenance and other logistical support, or do one of many tasks vital to the air force mission. The air force currently enlists about 5,000 male and female officers each year to fill openings in a wide variety of challenging careers.

Applicants for enlistment in the air force must be in good health, must be American citizens or registered aliens, and must attain the minimum scores on the Armed Services Vocational Aptitude Battery. They must also be at least eighteen years of age. Before taking the oath of enlistment, applicants may be guaranteed either training in a specific skill or an assignment within a specific area of ability. Applicants can enter the Delayed Entry Program and become members of the Air Force Inactive Reserve. Applicants agree to begin active duty on a specific date, and the air force agrees to train them in the skill area they have selected.

Marine Corps

The marines are a part of the Department of the Navy. The Marine Corps and the navy work hand in hand. Marines serve on U.S. Navy ships, protect naval bases, guard U.S. embassies, and provide a speedy strike force that is always prepared to protect U.S. interests anywhere in the world.

There are approximately 174,000 officers and enlisted marines serving in a wide range of

The United States Air Force is the aerospace branch of our nation's armed forces.

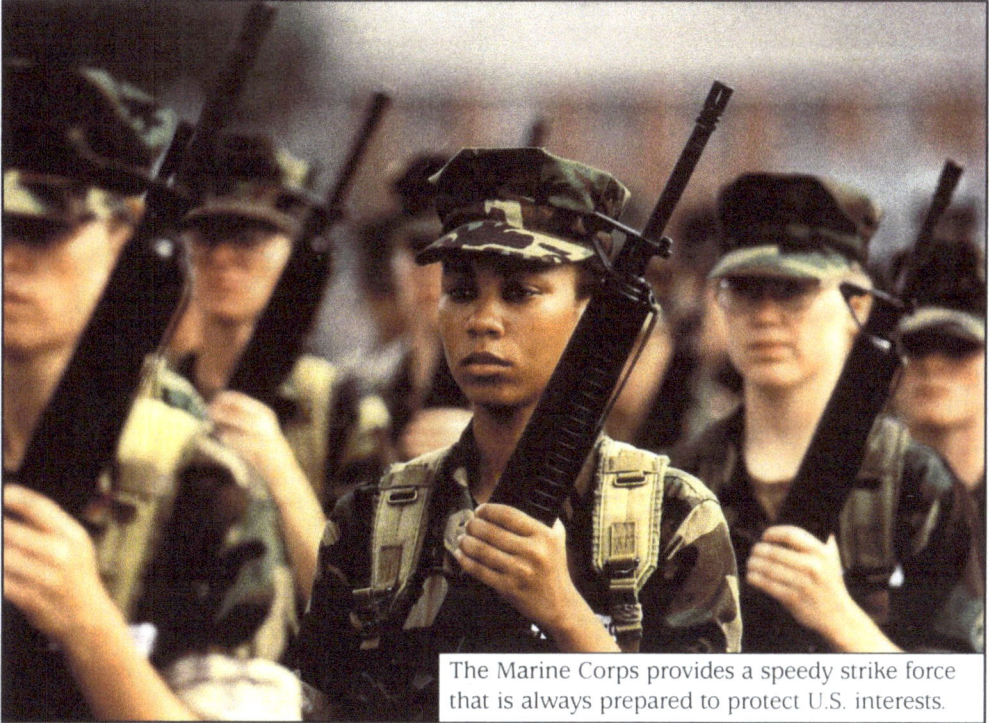
The Marine Corps provides a speedy strike force that is always prepared to protect U.S. interests.

positions. Some fly planes and helicopters, and some operate radar equipment. Others drive armored vehicles, gather intelligence, survey and map territory, and maintain and repair radios, computers, jeeps, trucks, tanks, and aircraft. The Marine Corps recruits approximately 41,000 enlisted men and women every year to fill openings in its many career fields. Marine Corps enlistees may sign on for three, four, or five years, depending on the type of enlistment program they select. Young men and women between the ages of seventeen and twenty-nine enlisting in the Marine Corps must meet demanding physical, mental, and moral standards. Applicants must be American citizens or registered aliens in good health to ensure that they can complete the strenuous physical training. The Marine Corps uses the Armed Services Vocational Aptitude

Among the many duties of the Coast Guard are the enforcement of customs and fishing laws and the regulation of boat and ship traffic in major harbors.

Battery to evaluate applicants' abilities. Applicants for enlistment can be assured a wide variety of options in training and duty assignments. Women are eligible for all occupational fields, with the exception of combat arms specialties such as infantry, artillery, and tank and amphibian tractor crew.

Coast Guard

The United States Coast Guard is crucial to American maritime safety. The Coast Guard saves lives and property in and around American waters. It enforces customs and fishing laws, protects marine wildlife, and fights pollution on our lakes and along the coastline. The Coast Guard watches traffic in major harbors and keeps shipping lanes open on ice-bound lakes. Equally important, the Coast Guard maintains lighthouses and other navigation aides.

The Coast Guard is a part of the U.S. Department of Transportation. In times of war, however, it may be placed under the command of the Department of the Navy. The Coast Guard has participated in every major American military struggle, even though it is the smallest branch of the armed services. At present, there are approximately 5,900 commissioned officers, 1,500 warrant officers, and 30,000 enlisted members. The Coast Guard has openings for about 300 new officers every year.

Individuals interested in joining the Coast Guard must be physically fit, must be American citizens or registered aliens, and must make at least the minimum required scores on the Armed Services Vocational Aptitude Battery. Enlistees in the Coast

Guard serve for four or six years of active duty. If openings are available, enlistees may be assured an assignment in a particular area of the country. Coast Guard recruits must be at least seventeen years old and must not have reached their twenty-eighth birthday on the day of enlistment.

Give It Some Thought

After talking to several people in his family and seeking the advice of several recruiters, Boris decided to enlist in the U.S. Marine Corps. He was surprised to discover how many career opportunities the marines could offer him. His ASVAB and physical were scheduled for the next week, then it would be only a matter of days before he would know for sure how he would spend the next few years of his life. After getting the best advice he could find, Boris was confident that he had made the right decision.

With five branches to choose from, either the U.S. Army, Navy, Air Force, Marine Corps, or Coast Guard could be the right choice for you. Be smart and take the time, like Boris did, to investigate your options by asking those close to you what they think you should do. But don't stop there. Stop in and talk to a few recruiters about your options. This chapter is just the beginning of your research. Read on to find out more about other military paths and career options.

3

Reserves

J uanita arrived to work on Monday morning feeling refreshed and ready to get back to her job as the general manager of a large department store. Many of her coworkers looked tired and unhappy. Most of them talked about what television shows they watched and how late they slept in on Sunday.

"What did you do this weekend?" Juanita's secretary, Brenda, asked.

"Oh, nothing much," Juanita replied with a smile. "I just repelled down the side of a cliff and built a bridge."

"You did what?" Brenda asked, astounded.

"That's the type of thing we do in the army reserves," Juanita replied.

There are two types of duty in the armed forces. The first is active duty, which means you are considered a full-time soldier. The second is reserve duty. Reservists are part-time soldiers and attend training only one weekend every month, plus two full weeks a year.

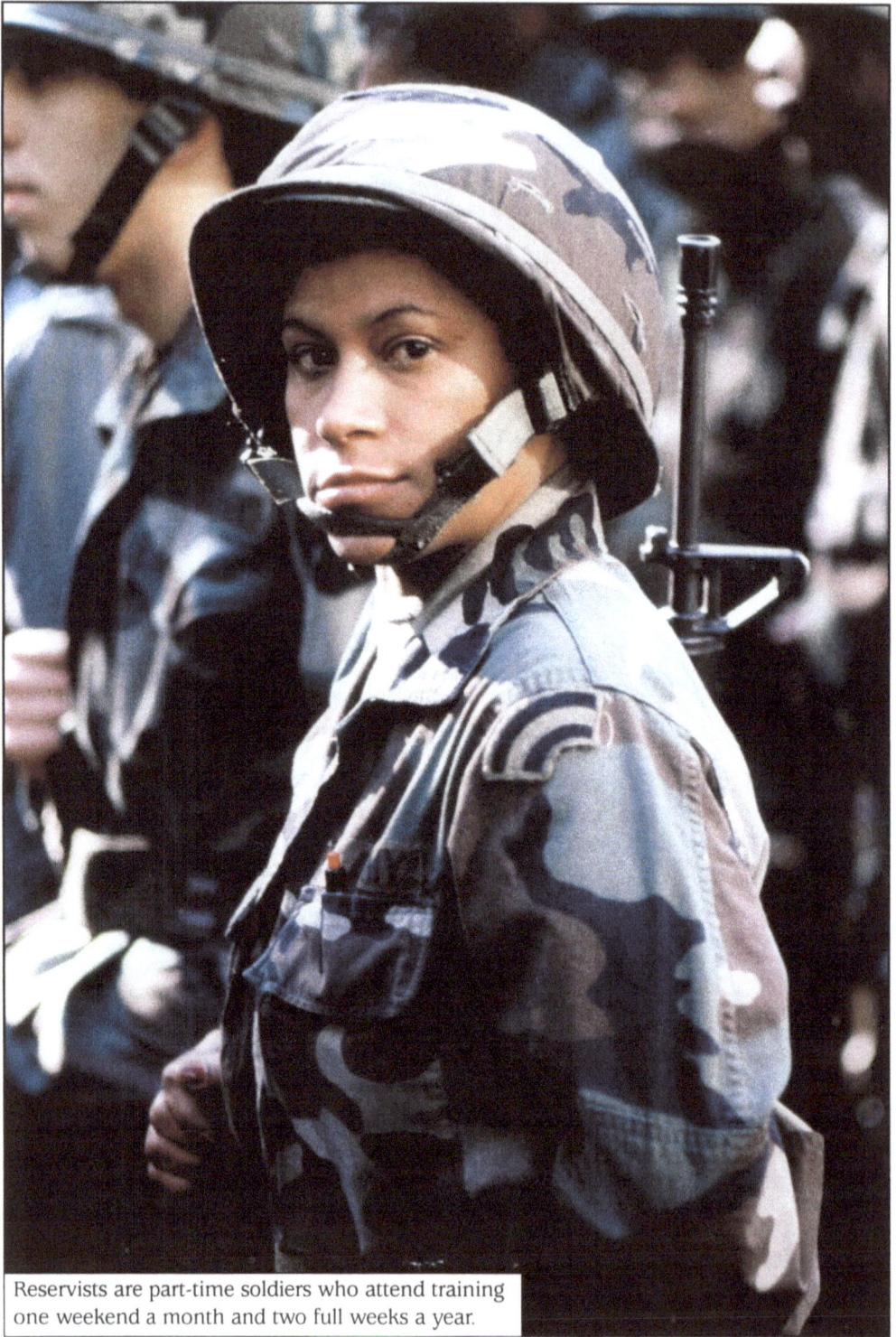

Reservists are part-time soldiers who attend training
one weekend a month and two full weeks a year.

There are several ways to become a reservist. First, people who enlist in a branch of the armed services have the option to serve part of their contract on active duty and the remainder on reserve duty. After serving a specified number of years learning valuable career skills on active duty, a person may decide to fulfill the remainder of his or her contract with the military through reserve duty while pursuing a civilian career at the same time. Second, enlistees may join the reserves directly. Individuals interested in taking this path must serve six months of active duty after basic training. These reservists are able to use the training they receive in the reserves outside the armed forces, immediately applying their new skills to civilian professions. Third, enlistees may enter the reserves as an officer after completing an ROTC program.

Once considered useful only for reinforcing military positions in the event of a national emergency, reservists today play a more involved role in the armed services. Whenever more active units are needed than are available, reservists are put in action.

For one weekend every month, these civilians gather at specified locations to train. The training is designed to duplicate situations that active-duty personnel may experience on a daily basis. In addition, reservist units spend two full weeks a year training as a team, perfecting the skills they have worked so hard to develop throughout the year, as well as working to learn new ones.

Much like active duty, reservist duty is considered a contractual obligation. Reservists must

31

Reservist training develops leadership skills that can be applied to civilian jobs and careers.

serve a specified period of time. They are paid for their participation in the armed forces and receive additional benefits as well. In fact, after twenty years of devoted service, reservists can receive the same retirement benefits as active personnel.

In addition to serving during times of war, two branches of the reserves—the Army National Guard and the Air Force National Guard—are often called upon to protect our country during national emergencies and disasters. These branches of the reserves are placed under the authority of state governments for the duration of their participation in the disaster.

Reservist training is valuable because it develops skills that can lead to civilian jobs and careers: airline pilots, police officers, computer programmers, even management positions can be earned with the leadership skills gained in the reserves.

4

More Than Just Soldiers

*R*oger and Ian had met in basic train-ing and had become close friends. They had just finished one full year in the U.S. Navy, and they were each enjoying the new positions they had recently been trained for. Roger was a trained compu-ter programmer, and his new assignment was aboard a destroyer, programming and troubleshooting problems in onboard computer systems. Ian was a newly trained naval cook and was overjoyed to have been granted a position as an assistant cook on the same destroyer as Roger. Ian was doing exactly what he always wanted to do, cooking for a living, and Roger had learned a new trade that would guarantee him steady work for the rest of his life. They were setting out for a six-month tour in the Mediterranean, and they were sure to have many new and exciting experiences to write home about.

The armed services offer enlistees a wide variety of vocational training opportunities.

Being a member of one of the five branches of the military means devoting yourself to serving your nation in times of war and peace. It means making a promise to your fellow Americans to defend them, their property, and their liberty when called upon to do so. Enlistees must first and foremost be interested in becoming a soldier.

The armed services, however, have much more to offer enlistees than a chance to become adept soldiers. Young men and women in the military are guaranteed solid, rewarding, hands-on training—provided they work hard and strive to do their best. Approximately 365,000 active-duty and reserve-duty soldiers and officers are trained by the U.S. military every year, making it one of the largest employers in the world. This chapter will provide you with twelve convenient vocation categories to consider as you plan for a future in the military.

Does the Military Have a Job for You?

Combat Specialty

Of course, the armed forces offer young enlistees first and foremost a chance to become professional soldiers. If you are interested in pursuing a lifetime career as an officer in the armed forces, refer to chapter 5. However, you don't have to be an officer to apply yourself to the area of combat specialty. Also, personnel skilled in combat specialties don't necessarily have to restrict themselves to a life in the military, as this area of expertise can open up a wide range of civilian opportunities after active duty is completed.

Combat engineers are needed in combat situations. Combat engineers combine combat skills and building skills to allow fighting forces to do their jobs with greater ease (for instance, rapidly building a bridge across a swiftly flowing river). Skilled infantrymen operate weapons and equipment. Special forces team members—like the army's Green Berets and the Navy SEALs—are necessary for offensive missions, demolitions, intelligence gathering, search-and-rescue missions, and other missions from aboard aircraft, helicopters, ships, or submarines. In addition to being trained for combat, special forces team members are often excellent swimmers, parachutists, and survival experts. Artillery personnel are taught to position, direct, and fire artillery guns to destroy enemy positions and aircraft. Artillery specialists are

The strong sense of discipline that one learns from training as a combat specialist can be indispensable in the civilian world.

usually trained in a particular type of weapon, such as cannons, howitzers, missiles, and rockets. Tank crews work as a team to run armored vehicles and fire weapons. Tank crew members generally specialize in a type of armor, such as tanks or amphibious assault vehicles.

While these skills are not always easily transferred to civilian careers (as in the case of tank training), the teamwork, leadership skills, and strong sense of discipline that is learned from these positions can be indispensable in the civilian world. However, some civilian career opportunities—such as construction, landscaping industries, forestry positions, demolition, police work, firefighting, lifeguarding or life-saving careers, diving, and swimming instruction—may seem like natural job choices for you after completing active duty as a combat specialist.

36

Human Services

If you like helping others, then you might want to consider a military career in human services. Enlistees involved in human services are trained to counsel other military personnel with social and/or substance abuse problems, such as drug addictions, alcoholism, depression, and other emotional problems. Human services specialists are trained to identify personal problems, interview personnel who need help, give advice to personnel and their family members, and teach classes on human relations. Training schedules include counseling techniques, psychological testing methods, and drug treatment procedures. After being discharged from the military, human services personnel will have many career paths to choose from, including counselors in rehabilitation centers, schools, colleges, hospitals, and other public organizations. It should be mentioned that some of these positions require a college degree in psychology or social work. Some human services professionals decide to become religious specialists and devote their careers to meeting the spiritual needs of military personnel. This also includes acting as a personal counselor should the need arise.

Media and Public Affairs

The modern military has a great need for individuals with creative abilities to fulfill various roles in the areas of communications and the arts. Photographers are needed for intelligence gathering and news reporting. Bilingual or multilingual enlistees can follow a career in translation, interpreting, or interrogating. Musicians in the armed forces perform at ceremonies, parades, concerts, and dances. Personnel trained in the areas of audiovisual technology, communications, and broadcasting create training films, record special ceremonies, and broadcast military events. Journalists and news writers help write and organize news programs, music programs, and radio shows for the armed services. Graphic designers and illustrators create military brochures, newspapers, training manuals, and posters, as well as television and movie special effects. Enlistees who get training in these areas are sure to have a great variety of jobs to choose from upon discharge, ranging from professional musicians to news reporters.

Health Care

Since the military supplies medical care to all men and women in the services, medical specialists are really important to the armed services. Medical and health care specialists give patients the attention necessary to help them recover from sickness or injury. Doctors, nurses, dentists, physical therapists, radiology technicians, pharmacists, optometrists, and laboratory technicians each play an important role in keeping the armed services running efficiently. Many of these positions can lead to successful civilian careers, although it should be mentioned that a college degree is often necessary.

Engineering, Science, and Technology

The armed forces are a perfect place to receive training in a scientific or technical field. Communications

Photographers are needed by the military for intelligence gathering and news reporting.

equipment operators synchronize air, sea, and ground forces. Space operations specialists run and repair spacecraft ground control command equipment, including electronic systems that track spacecraft. Weapons specialists transport, store, inspect, organize, and dispose of weapons and ammunition. Air traffic controllers direct aircraft into and out of military airfields, and track aircraft by radar. Weather specialists make weather forecasts to plan troop movements, airplane flights, and ship traffic. Computer programmers design and prepare computer programs that solve problems and organize data. Environmental safety specialists ensure that military facilities and food supplies are free from disease, germs, and other unsafe conditions.

This diverse field has many more roles that need to be filled, including mapping specialists, radar specialists, and intelligence specialists. Discharged individuals go on to be computer experts, CIA or national security employees, laboratory technicians, chemists, television meteorologists, and so on. Not all of these positions will require a college degree after the training you can receive in the military, although some might.

Administration

The five branches of the military are made up of numerous systems that need to work together fluidly in order for the branches themselves to run smoothly. As a result, individuals are needed to fill management and administrative positions on many levels. For instance, every year, trained male and female recruits who know the ins and outs of the

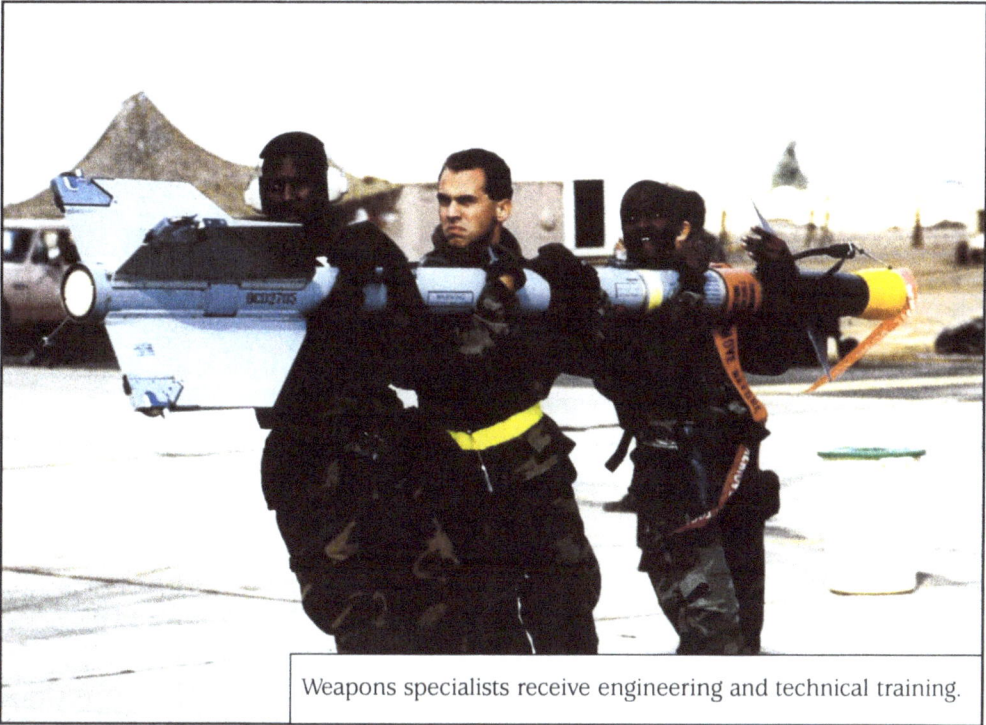
Weapons specialists receive engineering and technical training.

recruiting process are needed to attract new enlistees to the armed services. The armed services have their own postal service, and trained postal workers keep it running smoothly. Flight operations specialists keep the military's air fleet safe and proficient. Preventive maintenance workers keep all military vehicles and machinery up-to-date by organizing crews of maintenance workers.

The military needs legal specialists and court reporters to keep its court system functioning properly. Administrative support specialists record information, type reports, and maintain files to assist in the operation of military offices. Armed forces computer systems specialists make sure information is entered, stored, and processed economically and expediently. Transportation specialists organize and assist in air, sea, and land transportation for people and cargo, and some even work as gate agents and flight attendants. Military personnel trained to work in an administrative role can expect to find an abundance of career options open to them once they leave active duty. Those positions include general managers, travel agents, legal secretaries and aides, postal workers, finance and accounting specialists, and computer specialists, to name just a few.

Service

Much like the civilian world, the armed services have various service programs to help keep daily life and the general military organization running effectively. The military services have their own law enforcement and security workforce that investigates crimes committed on military property or that involve military personnel. They also guard inmates in military correctional institutions. The armed forces also have their own body of law enforcement—the military police—who monitor traffic, prevent crime, and respond to emergencies. Military firefighters put out and help prevent fires in buildings, aircraft, and aboard ships. The armed forces need food service specialists to prepare all types of food. They also order and inspect food supplies. Some military kitchens prepare thousands of meals at one time, while others arrange meals for small groups. Enlistees with military training in one of these areas can search for similar civilian jobs after being discharged. Many service-oriented personnel go on to become police officers, detectives, private investigators, undercover agents, correction officers, firefighters, security guards, cooks, chefs, bakers, or restaurant managers.

Vehicle and Machinery Mechanic

The five branches of the military rely heavily on an abundance of transportation vehicles and motorized equipment. Therefore, it is vital to have a staff of trained mechanics to keep tools and vehicles in tip-top shape. Automotive and equipment mechanics maintain and fix jeeps, cars, trucks, tanks, missile launchers, and other combat vehicles. They also maintain construction equipment. Aircraft mechanics check, tune up, and repair helicopters and airplanes. Skilled mechanics repair and maintain gasoline and diesel engines on an assortment of watercraft. They also repair mechanical and electrical equipment aboard ships. Heating and cooling mechanics install and repair air-conditioning, refrigeration, and heating equipment. Electrical and mechanical equipment in power-generating stations is installed, maintained, and repaired by a team of mechanics. Trained scuba divers repair underwater equipment. Military personnel trained in these areas can find civilian careers in some of the following areas: automotive repair; police and fire rescue units; underwater salvage companies; commercial airline mechanics; refrigeration, air conditioning and heating equipment specialists; commercial fishing; oil drilling operations; and utility company repair.

Electrical Equipment Repair

Specialists trained in electrical systems and electrical repair are always in high demand in the armed forces. Electrical specialists maintain and repair electric motors, electric tools, and medical equipment. The military relies heavily on

Specialists in electrical systems are always in demand in the armed services.

computer systems to run their day-to-day operations, and computer experts install, test, maintain, and repair these computers. Weapons technicians maintain and repair weapons used by combat forces. Electricians maintain and repair electrical systems on airplanes, helicopters, ships, and submarines. Other areas where electronic specialists help keep the military running smoothly include radar and sonar, communications, power plants, photography and media, and the repair and maintenance of precision measuring devices. In the civilian world, electronics experts can pursue careers as general electricians, computer equipment repairers and service technicians, electricians for aircraft manufacturers, and photographers and video systems maintenance workers.

Construction

Each year, the armed forces undertake an abundance of construction projects. In addition to this, specialists such as plumbers and electricians are needed to help produce safe structures on bases and other important military locations. Trained staffs are required to operate construction vehicles—like bulldozers, cranes, and graders—during the creation of roads, airfields, and buildings. Construction specialists work with engineers and other building specialists to assemble and repair buildings, bridges, foundations, dams, and bunkers. Plumbers and pipe fitters construct and maintain pipe systems for water, steam, gas, and waste. Building electricians install and repair electrical wiring systems in offices, airplane hangars, and other buildings on military bases. Military personnel trained in construction can look forward to many career opportunities after active duty. These include plumbers and electricians, heavy machinery operators, owners or managers of contracting firms, bricklayers, stonemasons, cement masons, and carpenters.

Machine Operator and Precision Work

There are a number of specialty occupations in the armed forces that require training in the operation of machinery unique to that area of work. Welders and metal workers make and install metal roofs, air ducts, vents, and gutters. They also make custom parts for repairing ships, submarines, aircraft, buildings, and equipment. Water and sewage treatment plant operators take care of systems that treat sewage and purify water. The armed forces count on experts to inspect, maintain, and repair survival equipment (parachutes, aircraft life-support equipment, air-sea rescue equipment). Dental and optical technicians make and repair dental fittings and eyeglasses for military personnel. Machinists operate lathes, drill presses, grinders, and other machine shop equipment when making and fixing metal parts for engines and other machines. The military produces many publications each year, and specialists are trained to operate printing presses and binding machines. Upon discharge from the armed forces, personnel trained as machine operators could find work as welders for a variety of companies (auto repair shops, aircraft manufacturers, construction companies), municipal public workers, dental and optical laboratory workers, commercial printers and print shop managers, or machinists in the heavy machinery industry.

Transportation and Material Handling

The transportation of machinery, cargo, supplies, and even personnel is a must in today's modern military. The armed forces put a great deal of effort into training individuals to specialize in moving and delivering important military assets. Drivers operate fuel or water tank trucks, tractor trailers, heavy troop transports, and passenger buses. A large team of pilots operate all types of aircraft in order to conduct intelligence missions, rescue personnel, transport troops and equipment, and perform long-range bombing missions. Aircraft launch experts supervise equipment used in aircraft carrier takeoff and landing operations.

The transportation and handling of machinery, cargo, and supplies is a crucial type of military training.

Flight engineers look over airplanes and helicopters before, during, and after flights in order to ensure safety. Boat operators pilot many types of small watercraft, including tugboats, gunboats, and barges. General naval personnel help run and maintain military ships, boats, and submarines. The military trains a large staff of cargo specialists to deliver provisions, weapons, gear, and mail to United States forces in many parts of the world by ship, truck, or airplane.

Since the military requires large amounts of petroleum products to keep its vehicles running, petroleum specialists are trained in the proper storage and shipment of oil, fuel, gases, and lubricants. Upon entering the civilian world after active duty, transportation experts will find themselves well prepared for careers as movers, bus drivers, fire truck drivers, pilots for commercial airlines and private shipping companies, industrial truck operators, longshoremen, cruise ship workers, and tugboat pilots, to name just a few.

Aren't you surprised to find that the armed services have so many great career opportunities to choose from? Working in the armed services gives you the opportunity to receive hands-on training while you proudly serve your country, make long-lasting friendships, see the world, and prepare for your future employment as a civilian. What more could you ask for?

5

You Could Become an Officer

*J*ulie was about to graduate high school, and most of her friends had been accepted to college. While college seemed somewhat tempting to her, Julie had chosen to hold off on enrolling for now. She wasn't sure she wanted to go back to school so soon. Senior year had gone well, but she was tired of just sitting in a stuffy classroom. Not that she minded going to school, but right now it just seemed to lack something truly appealing.

On the day after her graduation, Julie's grandfather, Mike, came by to congratulate her on ranking in the top 10 percent of her class. They had a long talk about Julie's future, and Mike had some very interesting advice to offer her. Mike had led a long, successful career as an officer in the U.S. Coast Guard and was lucky to have served very close to his home and family. The thought of following in her grandfather's

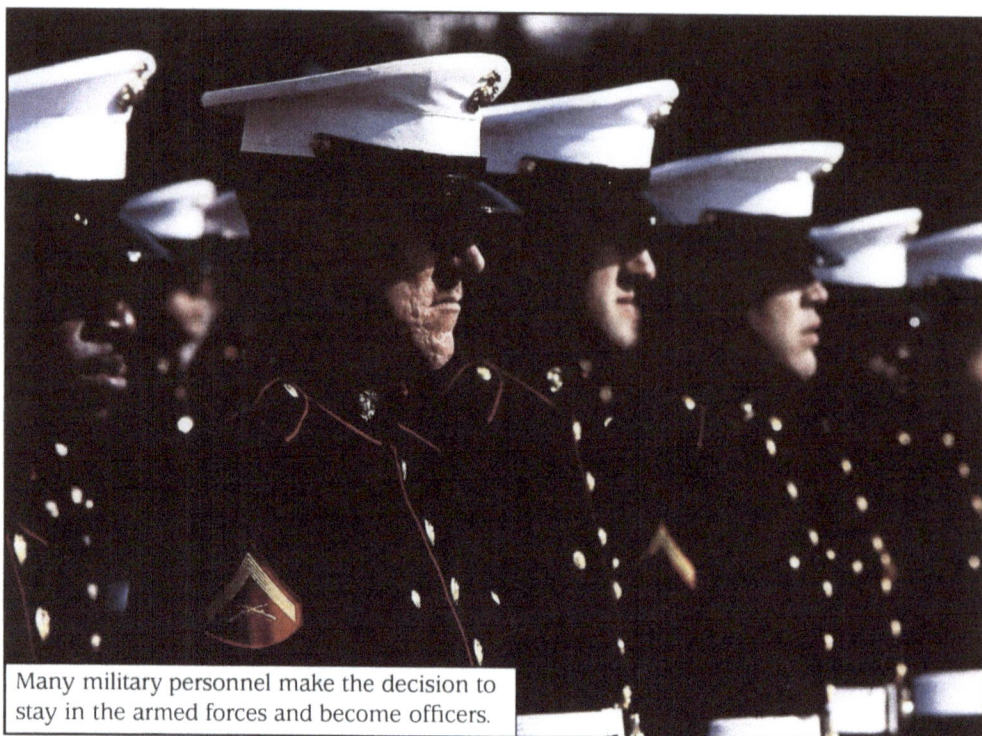
Many military personnel make the decision to stay in the armed forces and become officers.

footsteps made going to college seem much more enjoyable. To become an officer, Julie would need a college degree, or she would at least need to be working toward one. This was something Julie would seriously have to consider.

Many military personnel make the decision to become full-time soldiers instead of accepting an honorable discharge from the armed forces when their active or reserve duties have been completed. Becoming an officer is a career path in and of itself, and can result in a lifetime of fulfillment, not to mention the attractive benefits that are offered.

To become an officer in any of the five military branches, an individual must meet certain criteria. All branches require that the applicant be a high school graduate, pass a medical and physical exam, and be at least seventeen years old. In addition, to be considered for an officer position, an applicant needs to work toward or already have obtained a four-year college degree. If college is not in your future plans, becoming an officer may not be the right choice for you. If you do intend to earn an undergraduate degree, you may be the perfect candidate for officer status. As mentioned in chapter 1, the Reserve Officers Training Corps (ROTC) is another path you may decide to follow to become an officer in the armed forces.

Pursuing a full-time, long-term career in the armed services can be fulfilling and challenging. It certainly isn't the right choice for everyone, but it is as viable an option as the ones mentioned in the previous chapter. The military is always in need of effective leaders, teachers, and mentors to guide troops and individuals, and help them to become the best soldiers they can be. Officers are brave, quick-witted, and patriotic individuals. After twenty years of service, officers in the armed services are eligible to retire with a very attractive benefits package. The more you give to the military as an officer in the armed forces, the more you will get back from the experience.

It should be said, however, that this goes for any position you might hold in the armed services. You don't have to become an officer to get the most out of your military experience. No matter what career you choose to follow in the armed forces, the skills you learn and the memories you will take with you will last a lifetime. A career in the armed forces is a challenging, yet fulfilling path to take; reading this book is just the first step down that rewarding path.

Glossary

administrative   Referring to the individuals involved in the management of an institution or business.

aerospace   Relating to the science or technology of aircraft and flight.

amphibious   Able to function both on land and in water. Also a mission that involves naval forces landing on a shore via water.

artillery   Large guns or cannons operated by teams of soldiers, and the individuals who operate them.

civilian   Individual not presently enlisted in the armed forces.

demolition   The practice of destroying something with the aid of explosives.

drill instructor   Noncommissioned officer who leads new recruits through basic training.

honorable discharge   Discharge from the armed forces under favorable circumstances. A soldier who finishes his or her active duty is granted an honorable discharge. The opposite, a dishonorable discharge, is a forced discharge from the military for an offense unbecoming to an enlisted soldier.

infantry   Unit of soldiers trained to fight on foot.

logistical   In the military this refers to the process of obtaining, distributing, maintaining, and replacing materials and personnel.

maritime   Term used to designate anything related to marine shipping or navigation.

midshipman   Student training to become an officer in the U.S. Navy or Coast Guard.

multilingual   Ability to speak more than two languages.

radiology   Branch of medicine that uses radioactive substances to diagnose and treat certain diseases.

scuba   Portable tank containing compressed air used for breathing underwater. Scuba divers in the military use scuba gear for underwater missions. (Scuba stands for self-contained underwater breathing apparatus.)

warrant officer   Officer ranked between a noncommissioned officer and a commissioned officer. This position requires great skill in a particular area of technical expertise and the ability to teach that skill to others.

For More Information

In the United States

U.S. Army
(800) USA-ARMY

Army Web Sites
The U.S. Army Homepage
http://www.army.mil/

Association of the United States Army
http://www.ausa.org/

United States Army
http://members.aol.com/sapper1lt/index.html

U.S. Army Special Forces: The Green Berets
http://members.aol.com/armysof1/index.html

The Army Reserve
http://www.goarmyreserve.com/

U.S. Navy

Active duty: (800) USA-NAVY
Reserve duty: (800) USA-USNR

Navy Web Sites

The United States Navy
http://www.navy.mil/

United States Navy SEALs
http://www.sealchallenge.navy.mil/

United States Naval Reserve Force
http://www.navres.navy.mil/navresfor/

United States Naval Air Reserve Force
http://www.navres.navy.mil/navresfor/navair/

U.S. Air Force

Active duty: (800) 423-USAF
Reserve duty: (800) 257-1212

Air Force Web Sites

Air Force Link
http://www.af.mil/

Air Force Association
http://www.afa.org/

Air Force Reserve
http://www.afreserve.com

U.S. Marine Corps
(800) MARINES

Marine Corps Web Sites
Marines
http://www.marines.com

MarineLink
http://www.usmc.mil/

Marine Forces Reserve
http://www.mfr.usmc.mil/

U.S. Coast Guard
(800) GET-USCG

Coast Guard Web Sites
U.S. Coast Guard
http://www.uscg.mil/

Hilopilot's Complete U.S. Coast Guard Homepage
http://jupiter.spaceports.com/~uscg/

U.S. Coast Guard Academy
http://www.cga.edu/

U.S. Air National Guard
Phone: (800) TO-GO-ANG

Air National Guard Web Sites

Air National Guard
http://www.ang.af.mil/

Air National Guard Recruiting
http://www.goang.af.mil/home.asp

National Guard Association of the United States
http://www.ngaus.org/

U.S. Army National Guard

(800) GO-GUARD

Army National Guard Web Sites

National Guard Association of the United States
http://www.ngaus.org/

Army National Guard
http://www.arng.ngb.army.mil/

U.S. ROTC

(800) USA-ROTC

Air Force
(800) 522-0033

Navy Reserve Officer Training Corps
(800) USA-NAVY

ROTC Web Sites

Army ROTC
http://www.rotc.monroe.army.mil/information/

George Washington University Naval ROTC
http://www.gwu.edu/ ~ navyrotc/

Air Force ROTC
http://www.gwu.edu/ ~ afrotc/

Air Force Officer Accession and Training Schools
http://www.afoats.af.mil/

General

http://www.militaryinfo.com/
Large database of military information

http://www.armedforcescareers.com/
Armed forces careers

To contact your nearest recruiter, consult your local yellow pages under the heading "Recruiters." Your local recruiter can answer all of your questions about enlisting in a specific branch of the armed forces.

In Canada

Canadian National Defense Headquarters
Lower Library
2 North Tower
101 Colonel By Drive
Ottawa, ON K1A OK2
(888) 272-8207

Web Sites

D-Net: Canadian National Defense Web Site
http://www.dnd.ca/

Canada's Army
http://www.army.dnd.ca

Canada's Air Force
http://www.airforce.dnd.ca/

Canadian Forces Air Navigation School
http://www.cfans.com/

Canada's East Coast Navy
http://www.marlant.dnd.ca/

Canada's Pacific Naval Fleet
http://www.marpac.dnd.ca/

The Queen's Own Rifles of Canada
http://www.qor.com/

For Further Reading

DelVecchio, Valentine. *Cadet Gray: Your Guide to Military Schools, Military Colleges and Cadet Programs.* Morro Bay, CA: Reference Desk Books, 1990.

Dunbar, Robert E. *Guide to Military Careers.* Danbury, CT: Franklin Watts Inc., 1992.

Hutton, Donald B. *Guide to Military Careers.* Hauppauge, NY: Barron's Educational Series, Inc., 1998.

Italia, Bob. *Armed Forces.* Minneapolis, MN: ABDO Publishing Company, 1990.

Kappraff, Ronald M. *ASVAB Basics.* New York: Prentice Hall, 1994.

Learning Express Staff. *ASVAB Core Review: Just What You Need to Get into the Military.* New York: Learning Express, LLC, 1998.

Paradis, Adrian A. *Military Careers: A Guide to Military Occupations and Selected Military Career Paths.* Lincolnwood, IL: NTC Contemporary Publishing Company, 1999.

Thompson, Peter. *The Real Insider's Guide to Military Basic Training: A Recruit's Guide of Advice and Hints to Make it Through Boot Camp.* Upublish.com, 1998.

United States Department of Defense Staff. *America's Top Military Careers: The Official Guide to Occupations in the Armed Forces.* Indianapolis, IN: JIST Works, Incorporated, 1997.

Wiener, Solomon. *Everything You Need to Score High on the ASVAB.* Denver, CO: Arco Publishers, 1999.

Windrow, Martin. *A Concise Dictionary of Military Biography: The Careers and Campaigns of 200 of the Most Important Military Leaders.* New York: John Wiley & Sons, 1996.

Index

About the Author

Greg Roza is a writer and editor living near Buffalo, New York. In addition to working for a New York book publisher, he teaches poetry at SUNY Fredonia and edits a humor Web site. He has a master's degree in English from SUNY Fredonia.

Photo Credits

Cover by Corbis; p. 2, 22, 41, 47 © Pictor; p. 8 © John Neubauer/Pictor; pp. 12, 14, 20, 24, 30, 50 © Superstock; p. 17 © FPG International; p. 25 © David Doody/FPG International; p. 26 © Peter Gridley/FPG International; p. 32 © Mike Perry/Pictor; p. 34 © Tom Campbell/FPG International; p. 36 © Jeffrey Sylvester/FPG International; p. 39 © Adam Smith/FPG International; p. 44 © Bob Daemmrich/Pictor.

Layout

Geri Giordano

www.ingramcontent.com/pod-product-compliance
Lightning Source LLC
Chambersburg PA
CBHW050909210326
41597CB00002B/77